A LOOK AT WEATHER

CLOUDS

BY KATE MIKOLEY

Gareth Stevens
PUBLISHING

CRASHCOURSE

Please visit our website, www.garethstevens.com. For a free color catalog of all our high-quality books, call toll free 1-800-542-2595 or fax 1-877-542-2596.

Library of Congress Cataloging-in-Publication Data

Names: Mikoley, Kate, author.
Title: Clouds / Kate Mikoley.
Description: Buffalo, New York : Gareth Stevens Publishing, [2024] | Series: A look at weather | Includes bibliographical references and index.
Identifiers: LCCN 2022058440 (print) | LCCN 2022058441 (ebook) | ISBN 9781538287262 (library binding) | ISBN 9781538287255 (paperback) | ISBN 9781538287279 (ebook)
Subjects: LCSH: Clouds--Juvenile literature.
Classification: LCC QC921.35 .M54 2024 (print) | LCC QC921.35 (ebook) | DDC 551.57/6--dc23/eng20230126
LC record available at https://lccn.loc.gov/2022058440
LC ebook record available at https://lccn.loc.gov/2022058441

First Edition

Published in 2024 by
Gareth Stevens Publishing
2544 Clinton Street
Buffalo, NY 14224

Designer: Karen Quigley
Editor: Kate Mikoley

Photo credits: Cover, p. 1 Triff/Shutterstock.com; warm/cold front symbols NickJulia/Shutterstock.com; series art (cloud background) ekapol sirachainan/Shutterstock.com; series art (cartoon cloud) MimaCZ/Shutterstock.com; pp. 2-3 kzww/Shutterstock.com; p. 5 artjoao/Shutterstock.com; p. 7 Haris Djonko/Shutterstock.com; p. 9 Antrakt2/Shutterstock.com, ifong/Shutterstock.com; p. 11 ju_see/Shutterstock.com; p. 13 VectorMine/Shutterstock.com; p. 15 dugdax/Shutterstock.com; p. 17 MaleWitch/Shutterstock.com; p. 19 AdventureStock/Shutterstock.com; p. 21 GrumpyLivesHere/Shutterstock.com; p. 23 MarBom/Shutterstock.com; p. 25 dreamsky/Shutterstock.com; p. 27 J. Allen Frame/Shutterstock.com; p. 29 bubutu/Shutterstock.com; p. 31 OSweetNature/Shutterstock.com; pp. 31-32 kzww/Shutterstock.com.

Printed in the United States of America

CPSIA compliance information: Batch #CS24GS: For further information contact Gareth Stevens at 1-800-542-2595.

CONTENTS

Words in the glossary appear in **bold** type the first time they are used in the text.

FORMING CLOUDS

Clouds may look like big cotton balls or puffy cotton candy. But if you look at a cloud long enough, you'll see its shape change. That's because clouds are made up of millions of tiny bits of water or ice floating through the air!

Make the Grade

The tiny bits of ice that sometimes make up clouds are called ice crystals.

Water in gas form is called water vapor. There is always some water vapor in the air. Warm air can hold more water vapor than cool air. When air cools, some of the vapor **condenses** and forms tiny drops of water, called droplets.

Make the Grade

Water vapor is invisible, meaning it can't be seen. When it condenses, the droplets get big enough to be seen.

As air rises, its **temperature** cools down. This is why most clouds form when warm air rises. As more air rises, more droplets form. When enough droplets form, they come together to make a cloud.

Make the Grade

Water droplets form on tiny bits of things in the air, such as dust.

ALL KINDS OF CLOUDS

Scientists who study clouds put them into groups based on **characteristics** they share, such as how high they form and what they look like. The three basic types of clouds are cirrus, cumulus, and stratus.

Make the Grade

Sometimes, condensed water vapor stays near the ground. When this happens, it's called fog.

Scientists group clouds further. *Nimbus* means "rain" in an old language called Latin. There are nimbostratus and cumulonimbus clouds. These are kinds of rain clouds. *Alto* means "high" in Latin. Altostratus clouds are higher than stratus clouds. Altocumulus clouds are higher than cumulus clouds.

Cloud Levels

Make the Grade

Cirrostratus clouds share characteristics with both cirrus and stratus clouds. There are also cirrocumulus and stratocumulus clouds.

CIRRUS CLOUDS

Cirrus clouds are high in the sky. They're commonly more than 20,000 feet (6,100 m) up in the air. They look thin and **wispy**. Cirrus clouds look a bit like long streamers or ribbons in the sky.

Make the Grade

Cirrus clouds are often a sign that a change in weather is coming.

Cirrus clouds are made of ice crystals and commonly appear during nice weather. The shape of a cirrus cloud can help us tell which way the wind is blowing high up in the **atmosphere**.

Make the Grade

People sometimes call cirrus clouds "mares' tails." A mare is a female horse. The clouds got the nickname because some people think they look like a horse's tail.

CUMULUS CLOUDS

Cumulus clouds are the big, puffy clouds that may look like giant cotton balls. They often have flat bases with rounded tops. Cumulus clouds are fairly low to the ground and grow upward instead of spreading out wide.

Make the Grade

Cumulus clouds tend to mean good weather. However, cumulonimbus clouds, or **thunderstorm** clouds, form when cumulus clouds grow really tall.

STRATUS CLOUDS

Stratus clouds are low clouds that tend to spread out across the sky. They are commonly gray, **layered**, and look like fog. When stratus clouds are in the sky, there may also be light rain.

Make the Grade

A nimbostratus cloud is a layered cloud that makes lots of rain.

CLOUD COLORS

When you think of clouds, you probably think of the color white. The water droplets or ice crystals that form clouds **reflect** sunlight. When the different colors of the sun's **wavelengths** mix, we see them as the color white. This is why many clouds look white.

Make the Grade

On a nice day, you might see many bright white clouds in the sky.

When it is raining or about to rain, clouds may be dark gray. This happens because the water droplets **clump** together to form raindrops. This leaves more space between the drops of water and allows less light to be reflected.

Make the Grade

Clouds sometimes have dark bases but bright tops. This is because the bases do not get as much sunlight as the tops.

ON THE MOVE

If a cloud is made of water, why does it float? Each droplet of water is so small that it takes a long time to fall to the ground, like a piece of dust. Also, warm air rises and pushes up against clouds, helping them float.

Make
the Grade
A heavy cloud can weigh
more than 1 million pounds
(450,000 kg)!

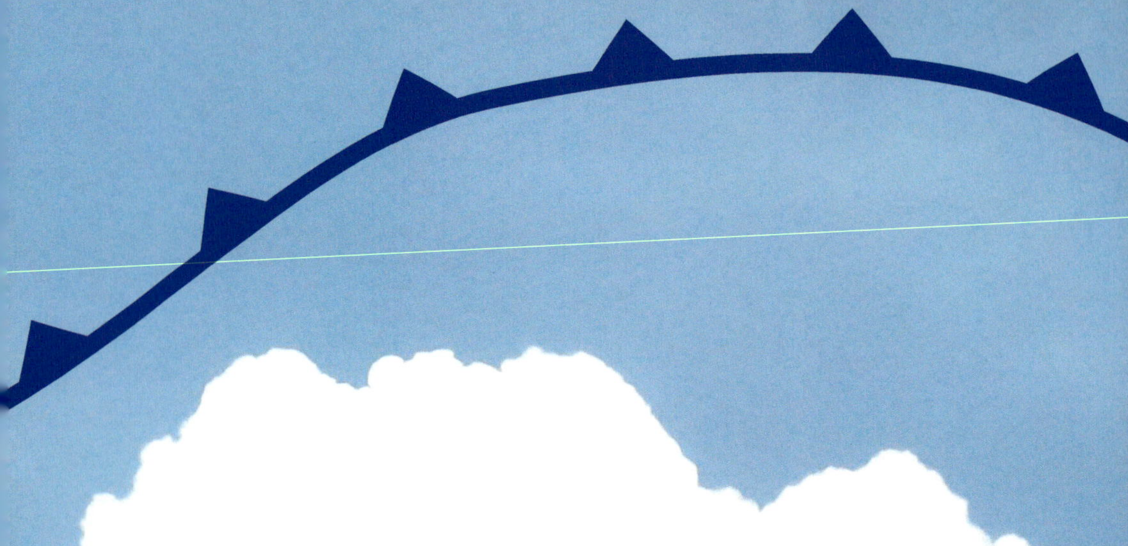

The next time you see a cloud, watch it for a minute. You will see it move! Cirrus clouds can move faster than 100 miles (160 km) per hour! In a thunderstorm, clouds often move at about 30 to 40 miles (48 to 64 km) per hour.

Make the Grade

Clouds aren't only in Earth's atmosphere. Other **planets**, such as Venus, Jupiter, and Saturn, have clouds in their atmospheres too!

FORMING RAIN IN THE CLOUDS

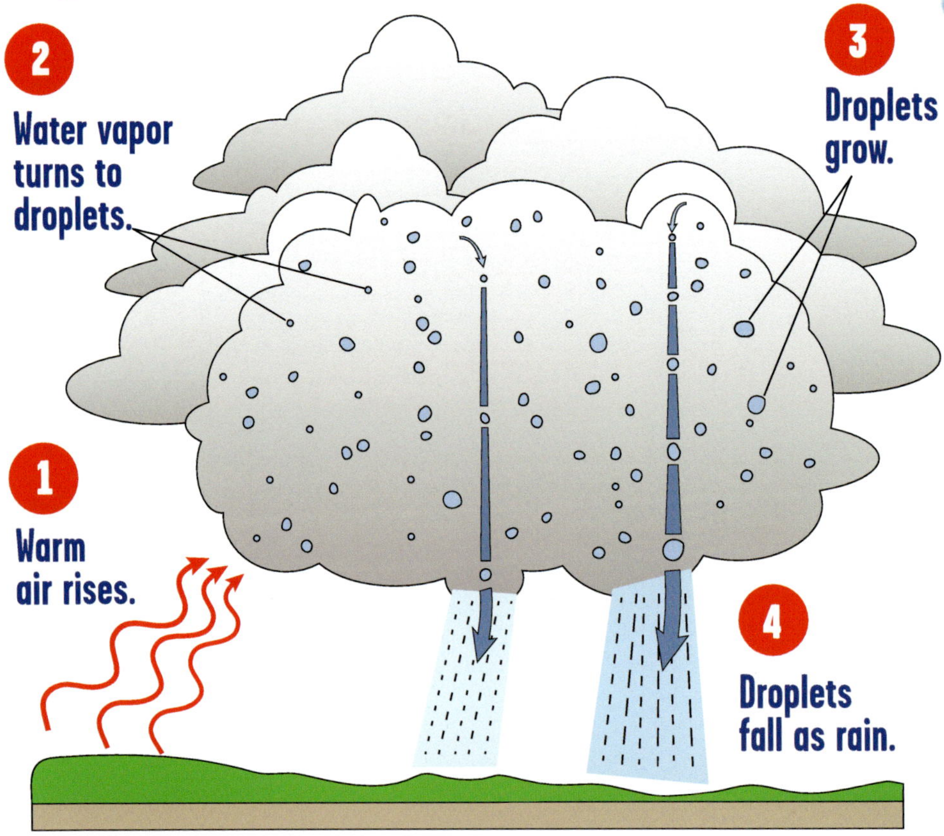

GLOSSARY

atmosphere: The mixture of gases that surround a planet.

characteristic: A quality that makes something different from other things.

clump: To form close together with other things.

condense: To change from a gas into a liquid.

layered: Having several thicknesses of something lying over or under another.

planet: Any of the large bodies that revolve around the sun in a solar system.

reflect: To give back light.

temperature: How hot or cold something is.

thunderstorm: A storm with thunder and lightning.

wavelength: The distance in the line of a wave from one point to the next.

wispy: Having the appearance of a thin, moving string.

FOR MORE INFORMATION

BOOKS

Horning, Nicole. *Let's Explore the Weather!* New York, NY: Cavendish Square Publishing, 2021.

Krajnik, Elizabeth. *Clouds and Precipitation*. New York, NY: PowerKids Press, 2019.

WEBSITE

How Do Clouds Form?
climatekids.nasa.gov/cloud-formation/
Check out this article from NASA to learn more about the formation of clouds.

Publisher's note to educators and parents: Our editors have carefully reviewed these websites to ensure that they are suitable for students. Many websites change frequently, however, and we cannot guarantee that a site's future contents will continue to meet our high standards of quality and educational value. Be advised that students should be closely supervised whenever they access the internet.

INDEX